CAREERS IN SCIENCE

What Do PALEONTOLOGISTS Do?

BENJAMIN PROUDFIT

PowerKiDS press.

NEW YORK

Published in 2022 by The Rosen Publishing Group, Inc.
29 East 21st Street, New York, NY 10010

First Edition

Portions of this work were originally authored by Lori Dowell and published as *Be a Paleontologist!* All new material in this edition was authored by Benjamin Proudfit.

Editor: Greg Roza
Book Design: Michael Flynn

Photo Credits: Cover (paleontologist), p. 21 Evgeny Haritonov/Shutterstock.com; (background texture) Anna Timoshenko/Shutterstock.com; p. 5 (man excavating) James L. Amos/The Image Bank Unreleased/Getty Images; p. 5 (illustration) Dorling Kindersley/Getty Images; p. 7 Time Life Pictures/Getty Images; p. 8 Underwood Archives/ Archive Photos/Getty Images; p. 9 shijie zhuang/Moment/Getty Images; p. 10 Digital_Lions/Shutterstock.com; p. 11 Carl De Souza/AFP/Getty Images; p. 13 Evgeny Haritonov/iStock/Getty Images; p. 15 Garry Adams/Photolibrary/Getty Images; p. 16 Lukiyanova Natalia frenta/Shutterstock.com; p. 17 Keneva Photography/Shutterstock.com; p. 19 Patrick Kovarik/AFP/Getty Images; p. 20 AFP/Stringer/Getty Images; p. 22 EvgeniyShkolenko/iStock/Getty Images; p. 23 picture alliance/Getty Images; p. 25 Pierre-Philippe Marcou/AFP/Getty Images; p. 26 Bettmann/Getty Images; p. 27 Michael Rosebrock/Shutterstock.com; p. 29 Superstudio/The Images Bank/Getty Images.

Library of Congress Cataloging-in-Publication Data

Names: Proudfit, Benjamin, author.
Title: What do paleontologists do? / Benjamin Proudfit.
Description: New York : PowerKids Press, [2022] | Series: Careers in
science | Includes index.
Identifiers: LCCN 2020040291 | ISBN 9781725329638 (library binding) | ISBN
9781725329614 (paperback) | ISBN 9781725329621 (6 pack)
Subjects: LCSH: Paleontologists--Juvenile literature. |
Paleontology--Vocational guidance--Juvenile literature.
Classification: LCC QE714.7 .P76 2022 | DDC 560.23--dc23
LC record available at https://lccn.loc.gov/2020040291

Manufactured in the United States of America

Some of the images in this book illustrate individuals who are models. The depictions do not imply actual situations or events.

CPSIA Compliance Information: Batch #CSPK22. For further information contact Rosen Publishing, New York, New York at 1-800-237-9932.

CONTENTS

TIME TO DISCOVER

How do we know what life on Earth was like millions of years ago? The animals and plants that lived in the past left behind a record of themselves! These are called fossils, and they include the remains of ancient plants and animals. Fossils can tell a lot about Earth, but it takes a special kind of scientist to find, study, and **interpret** fossils: a paleontologist!

Paleontology is the study of Earth's past through the fossil record. *Paleo* means old or ancient and having to do with the geological past. Paleontologists spend many years learning about the fossils that have already been discovered and what those fossils mean. Once at work in this awesome science career, paleontologists try to make discoveries of their own!

Fossils tell stories about how plants and animals grow and change, or evolve, over millions of years.

FORMING FOSSILS

Fossils are created when animals and plants or their **impressions** become quickly buried by **sediment**. Over many years, the buried sediment hardens. **Minerals** in the ground replace the shell or bone of the animal or fill in the impression. Fossils that are formed from the remains of a plant or animal, such as a bone or shell, are called body fossils. Fossils that record an action, such as a footprint, are known as trace fossils.

THE HISTORY OF PALEONTOLOGY

Interest in fossils and what they tell about Earth's history started long before scientists called this study "paleontology." Around 540 BC, ancient Greek thinker Xenophanes (zih-NAH-fuh-neez) noticed fossils of seashells on dry land. He believed that the land must have been underwater during an earlier period in Earth's history. In the eleventh century, Chinese scientist Shen Kua (SHUN KWAH) proposed an early theory of climate change when he discovered fossilized bamboo on land that was too dry for bamboo growth. He thought that the land in that area may have once received more rainfall.

By the 1700s, studying fossils had become a true science. During this time, scientists began to group kinds of fossils and record **details** about them.

Xenophanes studied philosophy, religion, and poetry as well as science.

WHAT'S A THEORY?

In science, theories are explanations for how something happens or why something works the way it does. They are ideas scientists can test based on what's been observed. Once a scientist has noticed something happening around them, they look for more information, or data. This data will lead them to make a hypothesis, or guess based on knowledge. Finally, they'll do more tests, or experiments. The outcomes of all these tests will inform the theory.

Georges Cuvier died in 1832. He spent much of his career working at the Museum of Natural History in Paris, France.

French scientist Georges Cuvier is called the father of paleontology. Based on his work **excavating** fossils and assembling the skeletons he found, as well as other scientific study, he convinced other scientists that animals can and have died out, or become extinct. In addition, he wrote a new system of **classifying** animals based on how their anatomy, or body parts, were arranged.

English scientist William Smith made another important contribution to early paleontology. He identified that rock strata, or layers, were always found in a certain order across England. Each layer had its own specific fossils too.

MARY ANNING

Mary Anning first began fossil hunting as a young girl. Around 1810, Anning and her brother found an *Ichthyosaurus*, an ancient marine reptile. She was only 12 years old at the time! Anning later found the whole skeleton of a *Plesiosaurus*. Because she was a woman, Anning didn't always get credit for her work. Even Georges Cuvier once said the *Plesiosaurus* fossil she found was fake. It wasn't!

Ichthyosaurus fossil

BECOME A PALEONTOLOGIST

In order to start a career in paleontology, one must earn at least a four-year college degree.

Most college students who plan to go into paleontology choose a biology major or geology major—or both! Geology is the study of Earth, especially of rocks, soil, and minerals and the processes they undergo. Biology is the study of life processes.

Since paleontology deals with remains of living things within Earth's geologic formations, it's easy to see why future paleontologists need to study both geology and biology.

Micropaleontologists and invertebrate paleontologists usually have a college degree in geology. Paleobotanists often major in a field of biology that deals with plants, called botany. Vertebrate paleontologists may focus on a field of biology called zoology, which is the study of animals. In addition, future paleontologists need to study other sciences, like chemistry and physics, as well as computers.

RESEARCH IT!

Most paleontologists also earn a master's or doctoral degree, which require more years of college. These studies are where paleontologists start to take more specialized classes, or those that have topics having to do directly with paleontology and the part of paleontology one wants to go into after school.

While earning higher degrees, future paleontologists work in labs. They begin to learn the **techniques** used for excavating and handling fossils. Paleontology students often do fieldwork with their teachers and put their knowledge to work. Near the end of their schooling, paleontology students complete **research** projects guided by a teacher and write a very long paper explaining what they've discovered. Often, this paper is published. After that, students choose their career path.

Fieldwork is work on a project outside of a classroom or lab done by a researcher.

KINDS OF PALEONTOLOGY

Paleontology is a big field of science. Those who study it specialize in one part of it, meaning they become an **expert** in one part of paleontology.

Animal paleontologists fall into two groups. Invertebrate paleontologists study animals without a backbone such as insects, corals, sponges, and starfish. Since these animals don't have bones, their fossils include fossilized shells and exoskeletons. Invertebrates also leave behind impressions of their bodies and tracks.

NOT JUST DINOSAURS

Dinosaur fossils are cool! But dinosaurs aren't the only animals that left behind fossils or the only fossils that paleontologists can learn from. Animal paleontologists may even specialize further than simply invertebrates or vertebrates. A vertebrate paleontologist may study fossils of the earliest fish. An invertebrate paleontologist may specialize in just fossilized corals—or even just one kind of coral!

Vertebrate paleontologists study animals such as fish, dinosaurs, birds, and mammals. They discover and put together the skeletons of animals from turtles to cats. Invertebrate and vertebrate paleontologists may work together since fossils of many kinds of animals may be found near each other.

These mastodon bones would be of great interest to a vertebrate paleontologist. Mastodons lived on Earth until about 11,000 years ago. They looked a bit like elephants.

Paleobotanists are paleontologists who study plant fossils from thousands or millions of years ago. These can include impressions of leaves or flowers. Paleobotanists also study fossilized wood, and even preserved plant parts like seeds. Paleobotanists also help study ancient environments, which is called paleoecology.

Micropaleontologists study microfossils, which are fossils that can only be seen with a microscope. These fossils may include those of algae, protists, and plant pollen. They may study the fossils of tiny, ancient crustaceans. Microfossils are often some of the oldest on Earth. That makes micropaleontology particularly interesting. Micropaleontologists play a big part in discovering what life was like billions of years ago on Earth!

Petrified Forest National Park in Arizona is full of petrified wood, or tree fossils!

WHERE DO THEY WORK?

Most paleontologists work at a college. They teach classes in geology, biology, and other sciences. They also do their own research, often with a team of students aiding them.

Many paleontologists work at museums that have fossils in their collections. They take care of the fossils in the museum and may use them in research. Museum paleontologists also work on how to display the fossils and ways to teach visitors about them. They may even lead talks about the fossils in the museum's collection.

Some paleontologists may work for the U.S. government or private companies. Their job may be to check for fossils at a work site and figure out what they are before work begins.

Paleontologists working at a museum have many duties, including putting dinosaur fossils on display!

DAY-TO-DAY WORK

Paleontologists are best known for their fieldwork. They find sites all over the world that may contain the kinds of fossils they're looking for. They spend time carefully digging into the rock and soil, noting the order of layers of rock and what is found in them. When a fossil is found, they use special tools to clean it off and bring it back to the lab or museum where they work.

dinosaur footprint

ALWAYS BE RECORDING

One of the most important things for paleontologists in the field to do is record everything about what they have found. From where in the rock strata a fossil was found to the fossils found nearby, this information could have a big impact on the meaning of a fossil discovery. Paleontologists take notes as well as many pictures of where they find a fossil so they don't miss a thing!

Sometimes, it can take months or years to fully excavate a certain site and find the whole skeleton or impression of a plant or animal.

Fossil hunting and discoveries are two of the most exciting parts of a career in paleontology. However, paleontologists also spend a lot of time researching and writing papers about their findings that they publish and share with other scientists.

Most paleontologists also work in laboratories. They study fossils closely to discover fossils' age, the climate conditions when the animals or plants lived, and other factors to tell a story about Earth as it once was.

Paleontologists use two basic methods of figuring out how old a fossil is. One is called relative dating. Fossils are found in sedimentary rock, which forms as layers, one on top of the other. Bottom layers are older than those above. So, if a fossil is found in a bottom layer, it's older than fossils found above.

Absolute dating is a way of finding a more exact date. Certain elements in rock break down at a constant rate. By measuring how much these elements have broken down, scientists can figure out how old something is.

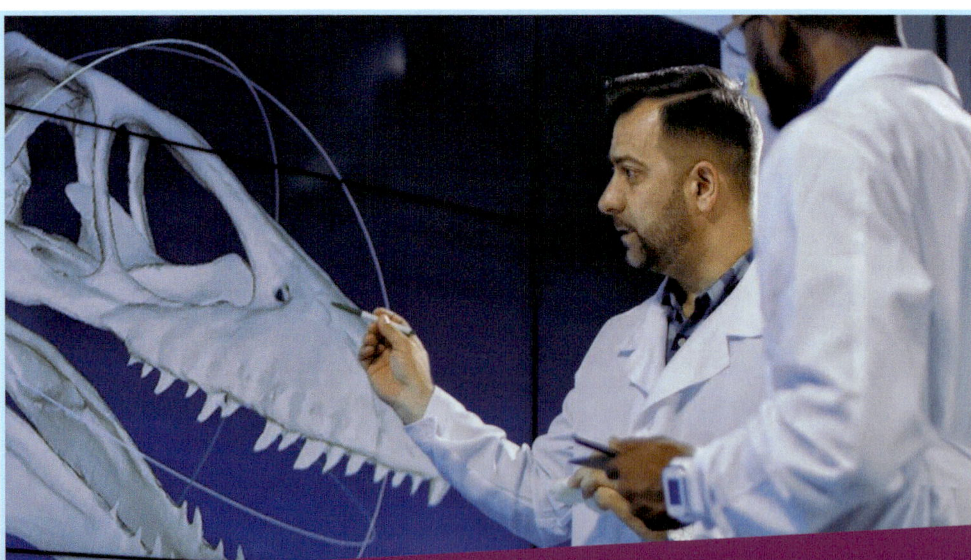

While a discovery in the field is exciting, discoveries are also made in the lab! Some scientists even prefer the close, careful work done in labs.

PALEONTOLOGY TOOLS

Sometimes paleontologists get lucky—a fossil may be **exposed** for anyone to see. However, they usually have to use tools to dig and dust off layers of dirt around fossils. They use trowels, or small shovels. Sieves, also called sifters, separate dirt and sand from around fossils. They use large rock picks, trowels, or drills to unearth fossils, and smaller picks and brushes for more gentle work. They don't want to destroy the fossils while trying to excavate them!

Back at the lab, paleontologists use modern **technology** to help them figure out more about fossils they find. They even have special computer programs to help look at data and create models of what animals might have looked like based on their fossils.

Whenever hammering hard rock, paleontologists wear safety glasses and gloves to protect their eyes and hands. They also carry magnifying glasses to help them to see smaller fossils.

UP CLOSE

An important tool paleontologists use to study fossils is an electron microscope. This microscope can show scientists the smallest parts of a fossil using **electrons** to magnify instead of just light. An electron microscope can magnify something about 1,000 times more than a microscope found in your school's science classroom!

BIG DISCOVERIES

Dinosaur fossils are still some of the most exciting discoveries paleontologists can make. Barnum Brown discovered the first *Tyrannosaurus rex* fossils in 1902. He became famous across the United States. When he traveled, people gathered at train stations just to see him step off the train!

In the Gobi Desert of Mongolia, Roy Chapman Andrews discovered the first fossilized dinosaur eggs in 1923. He also found new species of dinosaurs such as the *Velociraptor*. He's been called the "real Indiana Jones."

Sue Hendrickson, an amateur fossil hunter, found a complete skeleton of a *Tyrannosaurus rex* in South Dakota in 1990. You can see it at the Field Museum of Natural History in Chicago, Illinois.

Roy Chapman Andrews

In addition to exploring and finding fossils, Roy Chapman Andrews worked for the American Museum of Natural History in New York City for many years.

DINOSAUR NATIONAL MONUMENT

Dinosaur National Monument is located in the mountains between Colorado and Utah. Paleontologist Earl Douglass discovered fossils there in 1909. The remains date back 150 million years! "Dinosaur Wall" contains fossils of about 400 dinosaurs. More than1,500 bones remain in place but have been exposed for viewing through reliefing, which means paleontologists chipped away surrounding rock. It's thought that the dinosaurs didn't die here. Their remains were probably carried here by water.

YOU, AS A PALEONTOLOGIST!

Paleontology is a rewarding career in science. These scientists not only have a chance to make a fossil discovery of a previously unknown life-form, but also study it and try to explain how it fits into the history of Earth! Though the job takes many years of college to achieve, it's the perfect career for someone who likes science, enjoys being outside, and loves to solve mysteries!

Your path to a career in paleontology can start right away! Plan a visit to a nearby museum of natural history or even the Smithsonian National Museum of Natural History in Washington, D.C. Take earth science classes and join—or start—a fossil club at your school or community center. Someday, you might make the next big paleontological discovery!

Ask your science teacher if there are any places near your house where you can start looking for fossils!

GLOSSARY

classify: To assign something to a category or class based on shared qualities.

detail: A small part of something larger.

electron: A particle found in atoms that acts as a carrier of electricity in solids.

excavate: To unearth or carefully dig up.

expert: Someone who knows a great deal about something.

exposed: Not covered.

impression: A mark left as a result of applying pressure on something.

interpret: To tell the meaning of something in terms the audience can understand.

mineral: Matter in the ground that forms rocks.

research: Studying to find something new.

sediment: Matter, such as stones and sand, that is carried onto land or into the water by wind, water, or land movement.

technique: A certain way of doing something.

technology: Using science, engineering, and other industries to invent useful tools or to solve problems. Also, a machine, piece of equipment, or method created by technology.

FOR MORE INFORMATION

BOOKS

Crane, Cody. *All About Fossils: Discovering Dinosaurs and Other Clues to the Past.* New York, NY: Children's Press, an imprint of Scholastic Inc., 2021.

Skeers, Linda. *Dinosaur Lady: The Daring Discoveries of Mary Anning, the First Paleontologist.* Naperville, IL: Sourcebooks Explore, 2020.

WEBSITES

Paleontology: The Big Dig
www.amnh.org/explore/ology/paleontology
Learn much more about paleontology and discoveries from this interactive website.

Sally Walker—Paleontology
kidsciencechallenge.com/careers/paleontologist.php
Sally Walker, a professor of paleontology at the University of Georgia, offers more information about paleontology, including career options.

INDEX